Becoming a
Better
Builder

Winning in Construction Management

Stephen H. Dalton, AIA

outskirts
press

Becoming a Better Builder
Winning in Construction Management
All Rights Reserved.
Copyright © 2020 Stephen H. Dalton, AIA
v4.0

The opinions expressed in this manuscript are solely the opinions of the author and do not represent the opinions or thoughts of the publisher. The author has represented and warranted full ownership and/or legal right to publish all the materials in this book.

This book may not be reproduced, transmitted, or stored in whole or in part by any means, including graphic, electronic, or mechanical without the express written consent of the publisher except in the case of brief quotations embodied in critical articles and reviews.

Outskirts Press, Inc.
http://www.outskirtspress.com

ISBN: 978-1-9772-2800-0

Cover Photo © 2020 Dreamtine.com. All rights reserved - used with permission.

Outskirts Press and the "OP" logo are trademarks belonging to Outskirts Press, Inc.

PRINTED IN THE UNITED STATES OF AMERICA

Foreword

This book would not have been possible without my father, Cornelius Dalton, who taught me how to work with my hands on the many projects we did together when I was a child, and to my mentor, Artie Nusbaum, who taught me how to build seventy-story buildings in Manhattan faster and better than anyone else. Without both of them I would never have found my passion in construction management. I also want to thank my wife, Gloria, for putting up with me for two years as I wrote this book.

Stephen H. Dalton
Washington, DC
May 10, 2020

Table of Contents

Introduction i

Chapter 1 – The Difference Between Success and Failure—Be Afraid ... 1

Chapter 2 – Construction Is About Relationships
(It's All About People) 4

Chapter 3 – Don't Take it Personally 7

Chapter 4 – Do Take it Personally 10

Chapter 5 – Learn Something New Every Day 12

Chapter 6 – Remember Your Education and Training 14

Chapter 7 – Be Safe and Protect Others 17

Chapter 8 – Drink With Your Team 20

Chapter 9 – Curse as Much as You Can 24

Chapter 10 – Respect the Trades 26

Chapter 11 – Make Sure the Owner Likes You 29

Chapter 12 – Put in the Hours 32

Chapter 13 – Talk to People	35
Chapter 14 – Get Dirty	38
Chapter 15 – Have Integrity	40
Chapter 16 – Be Organized	43
Chapter 17 – The Dangers of Email	46
Chapter 18 – Don't Be Pushed Around	49
Chapter 19 – Support Your Team	52
Chapter 20 – Say Yes to Travel	55
Chapter 21 – Respect Your Mentors	58
Chapter 22 – Dealing with Idiots	62
Chapter 23 – Don't Forget Your Family	64
Chapter 24 – Visit the Factory	66
Chapter 25 – Promote Yourself	70
Chapter 26 – Don't Burn Bridges; Solve the Problem	72
Chapter 27 – Think Vertically and Horizontally	75
Chapter 28 – Persistence	80
Chapter 29 – Managing the Trades	82
Chapter 30 – Know When to Move On	85
Chapter 31 – How to Handle the Interview	87
Chapter 32 – Love What You Do or Find Another Career	91

Introduction

I was thirty years old standing on the sixty-fifth floor of a high-rise building on Fifty-seventh Street in Manhattan that was only a concrete skeleton with temporary railings. I looked down on the street below as the wind blew with such force, it felt like I would be lifted up off the floor. It was 1981 and construction in New York City had come out of a slump and was booming again.

I wondered how I had been lucky enough to be a project manager for one of the largest national construction firms at such a young age. Only a handful of people ever have the good fortune to end up with a job like this. At the time, I didn't know that I was part of a club that I would belong to for life. And those who had been building for many years instinctively knew who would make it in this club and who would not.

It started for me as an architect attending Columbia University in New York in 1975. After I graduated in 1978, the real estate and construction industry was in a terrible recession. I had been lucky enough to get a job for a medium-sized architectural firm and decided after three years of practicing and obtaining my architectural license that I wanted to get into construction. Only a small percentage of architects were accepted into the world of high-rise construction in Manhattan at that time. I was to be only one of three architects in a construction company of five hundred in 1981.

I had applied to several construction firms, but this was the one I wanted to work for. HRH Construction was the firm that was making the biggest impact on the skyline of New York City. I had managed to get an interview with the president of HRH and was quite nervous, as I knew little about construction at that time. I had arranged an interview with Artie Nusbaum, who was the president of HRH and a legend in New York, having built most of the high-rise skyscrapers in the city.

As I entered Artie's office, the first thing I noticed was his large size, as if he built these buildings with his own hands. He told me to sit down, went over my résumé briefly, and then said, "I don't like architects in construction. Most don't work out and they generally don't know how to build anything." I had so looked forward to this interview but was not expecting that reaction. To this day I don't know why I reacted the way I did, but I stood up and looked at him and said, "If you don't want me, then another company will, because you're not going to prevent me from doing what I know I love to do. So, I don't really give a shit if you hire me or not." Most likely the Irish side of me coming out.

As I started to walk out, he said, "Sit down." Surprised, I went back and sat down, as he leaned back from his large desk overlooking the streets of Manhattan from the 60th floor of this high rise on Third Avenue. Artie then said the words I always remember even after many years: "Well, at least you have balls. You start Monday." And that was the start of a lifelong relationship with Artie and construction. He died recently at ninety-two, and I told the story to his family at the service. They weren't surprised. He taught me how to build in the tough world of construction in New York. It was not the polite world of architecture that I had come from.

There are many books about construction management (CM) giving you a cookbook on what you need to know to be successful in building: scheduling, budgets, paperwork, quality control, contracts, project management software. But that's not only what makes you successful in construction. Success comes from who you are and what kind of person you are. It comes from a unique character that others in the field recognize.

To be a great builder for large complex projects as I have been involved with for over forty-five years, you need to have something special that those of us in construction recognize. It's a sense we get of who you are. It is what Artie saw in me when I walked out of his office.

Not everyone is able to be successful in construction. It's a rough and tumble world. Just because you go to a good college, study construction management, and read lots of books, it does not mean you're going to be a great builder. Yes, procedures and pushing paper are important—but not enough to ensure your success.

This book is a compilation of lessons that I have learned in forty-five years building large complex projects ranging from $10M to $250M. It is designed as a quick read with short chapters, as I know construction people have a limited attention span. You can skip ahead to chapters as you find topics that interest you. The order does not matter. This book is intended for those who are just entering construction and those who have practiced for a while.

I hope you learn something from my experience.

CHAPTER 1

The Difference Between Success and Failure—Be Afraid

They say the secret to getting ahead is not what you know but who you know. In construction, it's not what you know or who you know, but who you are, what kind of personality you have. I came into construction in an odd way, with a BA in psychology and then a master's in architecture. Most of my associates in construction were right out of engineering school as undergraduates. Technically very smart but not necessarily good builders.

With my background I understood people, both intuitively and from my studies in psychology. And then my master's in architecture from Columbia University gave me the technical skills to understand the intent of the designers, through their drawings, and the ability to construct a building that met their expectations.

But what I learned very quickly was that you succeeded or failed based on the fire in your belly, the determination to never fail, as well as your technical smarts and people skills. Today when I give advice to young engineers whom I work with, I tell them, "Be afraid." They look at me strangely because they have been taught to have confidence and not be afraid.

What I mean by "be afraid" is that you have to constantly think of everything that can go wrong when you are a construction manager or construction engineer on-site. You have to wake up in the middle of the night almost in a panic that you have forgotten something, or someone is not doing what they are supposed to do. It might be an architect; it might be a plumber or a factory making materials for your project. You have to think of every roadblock that could be thrown your way. You have to contemplate failure to succeed. And when you do, you will find there is very little you missed, and you are ready for the next day.

It's good to worry all the time, to think ahead and plan. If you do that, you will find that you get ahead without even trying. You will advance, get raises and promotions. Artie Nusbaum turned out to be my mentor and used to say to me, "Be the best at what you do, be the smartest, work harder, and the money and promotions will follow."

At the time, I thought you got ahead by complaining about your raise or bonus, complaining that you were not advancing in the firm. But quickly I knew he was right. As I worked harder, longer and smarter, I did get ahead, I did get promotions and new opportunities. And I did build faster than everyone else on the most amazing projects.

Another lesson I learned was to compete with other builders in the city. In New York City, we were always surrounded by other high-rises going up. My first building as a project manager was a sixty-five-story high-rise mixed-use building on Fifty-seventh Street in Manhattan, Metropolitan Tower.

At the time another similar high-rise was going up across from us on Fifty-sixth Street—a gleaming aluminum and glass skyscraper designed by another fancy architect. It was being built by a competing construction company and Artie would tell me to look across the street. If they are building faster than you then you are failing. Make it like a sprint to the finish line. Make it a competition.

And that's what I did. Every day I would look across the street, and if they were on the tenth floor with concrete and I was on the eighth floor, I would think of ways to build faster, to make the concrete contractor build smarter. And everyone wins that way, subcontractors and construction managers—and the subcontractors will thank you because they make money faster.

CHAPTER 2

Construction Is About Relationships (It's All About People)

For all the years I have been in architecture and building, my successes are a result of the relationships I have built. It may be on the project site, it may be at the architect or engineer's office, it may be an important vendor who is supplying marble or kitchen cabinets to your project, or it may be the owner. You need to talk to people.

I talk to people because I like people—the concrete foreman in the trenches of a sixty-five-story building, the project architect or engineer working on the project with you, or the owner's staff whom you deal with every day.

At Metropolitan Tower, one of my early construction projects, I did just that. No one told me to do it; it just came naturally. Today too many young engineers and project managers have been raised relying on the computer. They rely on it for emailing, texting, dating—everything. The person-to-person communication has been lost on several generations. In construction you can't build by email or texting. Sure, it's an important part of the building process. But without the personal relationships, you can't build. It's all about people.

CONSTRUCTION IS ABOUT RELATIONSHIPS (IT'S ALL ABOUT PEOPLE)

Harvey Mackay, in a book I really like called *Swim With the Sharks Without Being Eaten Alive*, talks about how to sell to clients. He has many sections on getting to know everything about a client in order to sell to them. He concludes that you don't sell by asking what they can do for you. You sell by asking what you can do for them. You ask for nothing and develop the relationship. It's the same in construction management. You first develop the relationships and then, when you actually need something, you just need to tell them the situation you are in and they will volunteer to help. Trust me, it works.

I remember once while building Trump Tower on Fifty-sixth Street and Fifth Avenue in Manhattan in 1982, I had a problem getting an electrician to go to an apartment and finish the work. Now, in this case it happened to be Steven Spielberg's apartment. Many rich and famous people had condominiums at Trump Tower, Donald Trump's first major project in New York City.

My boss, Aldo Rizzo, was a fiery redheaded man in his fifties with a temper to match and was a genius at working with people. I learned a lot from Aldo. He told me to get Spielberg's apartment finished, or he would kill me. Donald was on a tear and wanted it finished.

But at the time we were finishing the six stories of atrium retail as well as the fifteen floors of office space above it—heading toward a grand opening in February 1983. I was even working on the Trump Organization's offices on the 26th floor. So, the trades were stretched thin. But I had become friends with the electrical foreman of Lord Electric for no reason other than I liked him. We went out for drinks from time to time, and I helped him when he needed it.

I went to the foreman and begged him to send three men to Spielberg's apartment, and he did. The apartment had lighting, video, and sound systems that were way ahead of the time, and I needed electricians who knew what they were doing. My friend came through for me, and I was able to not disappoint Aldo.

Why are they there for you when you need them to do something? Because you're not just nice to them and talk to them when you need something. You already have established a relationship before that moment on the project when it is do or die and you need them.

Another chapter later on, called "Drink With Your Team," is an extension of this. Building should be the most fun you ever have in your career, and you should respect and like your team, both at the construction firm you work for, and with the trades, the architects, the engineers, the consultants, and the owner's staff.

Sure, there will be times when you want to strangle someone who doesn't provide manpower when you are on a tight schedule, doesn't return a shop drawing on time, or misleads you on a delivery time for materials. But you need to be firm without disliking them. Once someone thinks you dislike them, you are done, and they'll just wait for a time to pay you back for being a jerk. Don't be a jerk.

CHAPTER 3

Don't Take it Personally

Being sensitive is great in the theater or if you are a hairstylist or a psychologist. Being sensitive and having a thin skin in construction is a disaster. It can only lead to making mistakes and getting people angry at you. And email has made it worse. I remember one time I sent an email to an architect and I said, "I resent the email from the owner." He called me and asked me why I resented the owner and what had happened. I realized that I had actually taken an email from the owner and re-sent it to the architect. In other words, I had sent it again. It came out as *resent*. On a phone call or in person, that would not have happened. Misunderstanding happen all the time with email.

Another problem with email is responding too quickly. Before email you could take a deep breath before picking up the phone and calling someone to yell at them for a real or perceived action that got you really mad. And most likely even if you didn't take that pause, on the phone you would work it out. Now with email people tend not to take a breath. My mother used to say, "Count to ten before you react." Having had a bad temper as a kid, that was not easy for me.

Nowadays when you get an email that gets you riled up, you tend to immediately send a response and then push send. How many times have you sent a text or email and then said to yourself, "Maybe I should have waited"? But you can't get it back. I don't care what Microsoft says about "recall an email." When I see that someone has recalled an email, I read it more carefully because I know they screwed up. Sleep on it before you react if you are upset. Remember the result you want, not just making yourself feel better.

Now, I am not saying just take it on the chin. Just the opposite. If someone is not responding on time to build your project and you really need them to react, you have every right to fight back. But be smart about it, not emotional. The subcontractor or vendor who screwed you with a delivery or low manpower is not attacking you personally. They are doing what is best for themselves and their firm.

You need to straighten them out and make them realize that your project has to take center stage, not someone else's. And you don't do that by reacting emotionally. You do that through cold calculation. Don't get mad, get even. That's the Irish phrase that I like, and you have time to do that. You can remind them later what they did and make your point when they need you. They won't do it again.

I'll give you an example. I had a carpentry contractor once on a high-rise building in New York City who would not give me enough men to keep up with the schedule. I sent him the usual letters citing the contract, but he ignored them. I called him and he didn't pick up the phone. Now, I knew the owner of the company pretty well. His monthly check was due at that time, and one way to get the attention of a contractor is to not pay them. But that can result in other problems, like giving him an excuse to delay the project.

So, I sent him a fax of the check for $750,000 and said it was in my office and he could have it if he gave me the twenty men I wanted. The next day I had twenty men on the project. But I did it without emotion. I just casually told him he wouldn't get his check. No emotion and no personal attacks by me. That is much scarier to a subcontractor than yelling and screaming. Respect for everyone—but accountability. Everyone has to do their job.

CHAPTER 4

Do Take it Personally

I know, you think I'm crazy. Why would I have two chapters that seem to be opposite of each other? Because there are times when you should take what is happening on your project personally. Artie Nusbaum, whom I mentioned before, used to tell me that I should see building as a race, and anyone or anything that got in my way should be moved out of my way. I had to sweep the road clear, metaphorically, so all the trades could follow one after the other in sequence. Never let one trade block another, because it holds up everything. When I first started in construction, I was too calm, too patient. That drove Artie nuts.

So, he would tell me that when a subcontractor didn't do what they were supposed to do and delayed me, it was like they were punching me in the face. He then asked me if I would be upset if someone punched me in the face.

I said of course I would. Artie taught me to take it personally when anything got in the way of finishing on time or someone charged me too much for something. You need to be firm, be aggressive with architects, engineers, subcontractors, and vendors who are not moving fast enough. But if you do it with resolve and not emotion, you will get responses from people.

I used to call subs sometimes ten times a day if they would not respond the way I wanted. One time I sat for three hours in an architect's office until the architect came back from a meeting. He hadn't returned a marble shop drawing on time in spite of promising me the drawing. I got in a taxi, went to his office, and waited for him to return.

They thought I was crazy in his office, but when the architect returned and found that I had been waiting for three hours, he went right to his desk and marked up the shop drawing in front of me. I left with the approved drawing and sent it that night FedEx to the vendor who could then start fabricating the stone. Had I not done that, it would have been another week delayed in production. You need to know when to go crazy and when to leave it alone. That can't be taught in school. It comes from instinct.

CHAPTER 5

Learn Something New Every Day

If you ever get to the point where you think you know everything in construction, retire or find another profession. Artie Nusbaum used to have an expression that only a few of us understood. If he was on the jobsite and he saw a project manager or superintendent who thought he knew it all or had no energy or knack for construction, he would say, "It's not too late to go into hairstyling" and then walk away.

The person it was directed at would look baffled and might later ask me what it meant. My answer was simple—look for another job because he has watched you, and you don't have what it takes.

Artie had an incredible instinct for building and for people who could or could not build. He didn't need psychological tests, multiple interviews, long lists of references. I went through many of those later in my career—Harvard Management 101-type interviews and psychological tests. In construction we meet with you for thirty minutes and know if you can build or not.

One story I remember from Artie was about an engineer we were working with. By now I was five years in the company and had a good relationship with him. He had learned that, even though I was an architect whom he had little respect for as a profession, I could build.

On this day I complained about an engineer who had thirty years' experience and seemed to be an idiot and couldn't solve problems. I wondered out loud how you could be in a profession for that long and still not know anything. Artie corrected me and said, "You're wrong. This engineer doesn't have thirty years of experience. He has one year of experience repeated twenty-nine times." I never forgot that. This engineer started out thinking he knew everything, and he had nothing new to learn. Those are the ones that you should be afraid of since they make life-and-death engineering decisions each day.

I have never been on a project where I didn't learn something new. Whether it was the first four-sided, structurally glazed curtain wall system at Metropolitan Tower or deep foundations into rock on a seventy-story high-rise, I have always found new experiences to learn from. The fun is learning something new that you take with you for the rest of your career. Then forty years later when you sit in a room with architects and engineers, they will never be able to present you with a challenge that is too complicated. You take a piece from every project and put it together to solve the problem. It's a high that only those in construction understand.

CHAPTER 6

Remember Your Education and Training

I don't want you to go away thinking that my take on construction is that you just "shoot from the hip" and use your instincts only. What you learn in school is important, whether it's engineering, architecture, or construction management. Listen to your smart teachers, study hard, and take good notes. Keep a journal or a sketchbook.

You will find as you go on in your career that you keep coming back to the basics. How to read drawings, how to interpret specifications, how to keep accurate logs and records, how to perform a calculation, how to work with software programs. There's no substitute for systems that have been used for years.

When I started in construction in 1981, there was no computer in the office, no email, no cell phones. Yes, as my daughters say, I am very old. In 1984, we had just gotten fax machines in the office which we thought was really cool.

But somehow, we still built high-rise buildings on the crowded streets of Manhattan. How did we do it? Old-fashioned paper logs and records.

REMEMBER YOUR EDUCATION AND TRAINING

My first major project, Metropolitan Tower in New York City, was a mixed-use, sixty-five-story building on Fifty-seventh Street right near Carnegie Hall.

We had a good plan clerk, Alfonso, the guy whose job it was to send and receive drawings and documents to architects, engineers, contractors, subcontractors, and the owner. He was well-organized, and that was his full-time job—logging everything coming in and out and making sure things got delivered to the right person on time. My other staff—superintendents, assistant superintendents, project managers, assistant project managers, and administrators—relied on him to handle everything. No computers.

And Alfonso kept us organized. He was the lowest paid guy on the project but really the most important. Anytime I asked him to find a record or copy of something that was sent out, he found it in five minutes. We had rows of file cabinets with drawings and papers, and he knew his system and had an amazing memory.

My point is that he stayed with the basics, did not try to reinvent the wheel, and it worked. We built this high-rise in the middle of a busy city on Fifty-seventh Street in less than three years. And the supers and project managers did not have iPads to carry around, did not have cell phones with email. They studied the drawings, did their homework on the project, and had to remember the details of the drawings. Sure, they would go back to the office and check drawings or specifications, but they knew the system and they stayed with it.

I remember many of the lessons I learned at Columbia University while getting my master's in architecture. One that particularly sticks with me is advice from my structural engineering professor. He made us all promise to never design anything outside our area of training and expertise. Don't think you can size a beam or do anything that involves structure without a structural engineer. Stay in your lane.

When it comes to construction management technology today, it can be a great tool. But remember that there is no substitute for remembering the basics.

CHAPTER 7

Be Safe and Protect Others

All of my good bosses in construction were sticklers for safety. Artie Nusbaum was a prime example. He had grown up in the industry, was an engineer, had been in the Navy Civil Engineering Corps, the Seabees. And he was the general superintendent for the largest New York construction firm, Morse Diesel, in the 1950s and 1960s. His job entailed going around to all the construction sites and tracking every superintendent and assistant superintendent on all projects, probably three hundred people. And he was responsible for the on-time construction of everything.

He taught me that safety on the job is more important than building fast. He had seen many people injured or killed as I came to see later in my career, and it made him watch for safety protocol on all his projects. All construction workers have this in common—we've seen too many people get injured because it's a dangerous business, and we don't want anyone to get hurt.

We all took safety courses, what we call OSHA 30, and other training courses. Yes, it is boring and takes time away from what we like—building. But you have to pay attention to safety on the job. Safety is everyone's responsibility. You know the expression today, "If you see something, say something." Well, in construction I would call it "If you see something, do something" when it comes to safety. Don't make notes and put it off for later. Stop what you are doing and make it safe. In the time it takes you to look at your notes later, someone could be hurt or die. If you see, like I did one time, a barricade missing at an open elevator shaft, you go get the carpenter foreman or the nearest carpenter and you wait while they fix it.

You would think that the carpenter might be annoyed and not want to stop and fix it. But I've found that all workers understand safety because they have all seen fellow workers hurt. Every time I asked someone to fix a safety issue, they stopped what they were doing and did it. Period.

And construction is a band of brothers. Construction workers all react quickly in an emergency. I remember on Metropolitan Tower once a worker on the forty-second floor cut his finger off with a saw. It happens more than you would think in construction.

I was a few floors away and heard about it on my walkie talkie (no cell phones then). I rushed up to the floor a few minutes later. By the time I got there, ten workers were with the man who cut his finger off, and they had stopped the bleeding. The ambulance had already been called. One worker took the finger, ran down forty-two stories to the street, went to the nearest deli, and got ice to preserve the finger. By the time the paramedics got to the job, the workers had the situation under control and handed them the finger.

The finger was later attached at the hospital in surgery thanks to the quick thinking of that worker. And his fellow workers either went to the hospital later or asked how he was doing. Construction workers care about each other and care about safety.

Another part of the story that I remember reflects how close the construction team is. What I didn't tell you is that when that worker went down to get ice at the deli, the owner told him it would cost $2 for the ice. The worker explained that it was for the finger, which he had with him. The owner didn't care and charged him $2. Now, all the workers ate in that deli every day because we had a crew of five hundred on the job.

When the workers heard what happened, about a hundred of them went down to the deli, filled the store up so you couldn't even move, and told the owner if he didn't return the $2, they would boycott the deli, picket the deli, tell everyone what happened, and shut him down. I actually cleaned that up, because I don't want anyone to go to jail. Don't mess with the brotherhood of construction workers. And yes, the owner returned the $2. It was a matter of principle for all of us.

CHAPTER 8

Drink With Your Team

Construction workers drink. That's just a fact. Are they alcoholics? No (well, maybe). You know the old joke: you're not an alcoholic if you don't go to meetings.

I would say construction workers and those in the construction world are good recreational drinkers. And going out for drinks after work with your construction management team, your architects, your engineers, and the construction workers is a special type of bonding that brings you closer. If you don't drink, we don't trust you. Maybe it was the Italian and Irish trades in Manhattan at the time that makes me think that.

Construction people are funny. They have great stories, they work hard but like to have fun, and it's a real compliment for a construction foreman to ask you out for a drink. It means he accepts you as part of the team, not just an annoying pest who keeps asking him for things. Without going out socially after work and drinking—or, like we did in the old days, drinking at lunch—you don't have the same bond. That is the world we work in.

Speaking of drinking at lunch, I remember I was on a project at 750 Lexington Avenue in New York. It was a complicated office building designed by Murphy Jahn Architects right across the street from Bloomingdale's. I got on the project when the steel was about at the third floor.

The ironworkers on this project were mostly Native Americans. They were often working in structural steel because they were fearless walking the steel beams and erecting high-rise buildings. And I learned that my ironworkers drank beer every day at lunch.

Now, if you're not in the business you are saying to yourself, "They have one of the most dangerous jobs in construction, walking on steel beams sometimes sixty stories in the air in the wind, and they drink at lunch." Well, yes, they used to.

At this phase of the project, the steel erection schedule was the most important thing on the job. If the steel didn't go up on schedule, the whole project suffered. As I got to know the ironworkers one day, they said, "Come to lunch with us." I had heard they drank at lunch but had never been with them.

I went across the street to their favorite pub and had lunch. Well, six beers later I weaved out of the pub and was not very effective in my office the rest of the day. These guys acted as if they had not even had a drink, went back up sixty stories, and kept working. I'm not advocating this for anyone, but it shows how important the pub scene was in New York at the time for them to consider you part of the club.

And once you're in the club, they will do anything for you. They want to succeed themselves because it's a competition for them too. And they make more money the faster they build. And they want you to succeed.

The same goes for architects and engineers, although here for some reason we call it "Happy Hour." Make sure you take your design team out regularly for drinks. It makes a big difference.

As for the owner, most of the principals won't ask you for drinks because you're not their peers. But their staff is another story. Every owner has the front office staff who you talk to every day. You do need to get to know them, and drinks are one way, because they are the ones who will approve the monthly pay applications that keep the project funded and process your change orders. And a personal relationship is the only way to get things done.

One funny story—an example of the wrong way to use drinking to socialize—happened at Metropolitan Tower. It was a Friday and my team had completed a really difficult week. It was a late lunch and we decided to go the Spanish restaurant around the corner. Well, eight pitchers of sangria later, we were having fun and thinking we would just take taxis home and the week was done.

But as my project manager and I returned to the site to get our bags, Harry Macklowe, the owner of Metropolitan Tower, was just pulling up for a sales meeting at the condo office in the new building that was still under construction. Harry was a smart, educated, fun guy but this was bad timing. He saw us and said, "Please come to the sales meeting. We have a few questions for you guys." This was really bad. We knew that we were loaded and could not predict what would happen.

To make matters worse they were drinking vodka at the sales meeting to celebrate a sale of one of the condos, and of course, we had another drink. That was the last thing I remembered. I don't know how I got home, but the next day I woke up and remembered saying to Harry that night, "You know, Harry, the problem with this project is that your architects are idiots and your staff is not helpful when I need them."

Not knowing if I still had a job, I called the vice president over at Harry's office and asked him how much trouble I was in. Luckily, he knew me well and told me that Harry thought I was funny and actually agreed with me on my drunken assessment of the project. But I was lucky and would not recommend repeating my adventure.

CHAPTER 9

Curse as Much as You Can

As soon as I started working in construction, I learned how important cursing is. It's part of the construction language, and you have no credibility if you don't curse. I came from architecture, where if you started cursing in the middle of the drafting room, everyone would stop and wonder what was wrong. My mother taught me not to curse, so it was alien to me.

On my first construction project just out of the architectural firm, I would talk politely to a foreman, and finally one day the concrete foreman, Sammy Ianotta, said to me, "Kid, you have to start using the F-word. It's eff this and eff that. It's a verb, an adjective, a noun. It needs to be every third word."

He went on to explain that workers would not take me seriously if I didn't curse. So, I started to curse and it felt good. I would greet the labor foreman in the morning by saying, "Jack, how the eff are you. What the eff did you do last night? I need you to get this effing floor finished." It was liberating and I loved it.

In the field office everyone cursed. The F-word was cleverly woven into conversations in ways I never could have imagined. It was *Get me an effing cup of coffee. This guy is un-effing-believable. Eff this asshole.* It was as if I had been waiting my whole life for this.

The trick became turning it off when you left the office. I had two small daughters at the time, two and five years old. I was pretty good at turning it off at home unless I got a work call. Then it would turn back on.

I remember once my older daughter, Abby, was fighting with her mother. She was maybe five or six and she was very frustrated with what her mother told her to do. She went into her bedroom, thought for a minute, and then came back into the living room, looked straight at her mother with her hands on her hips, and said, "Eff you." And then she went back into her bedroom.

It was amazing. Her syntax, timing, and use of the word were perfect. There was nothing else to say. Of course, my wife looked at me and said, "Deal with that." Yes, I had used so many curses when I got work calls at home that I was oblivious to the fact that my daughter was standing next to me.

I went into the bedroom and tried to explain that we don't use words like that. My daughter, being very clever, smiled and said, "But, Daddy, you use the F-word." Since I was the adult, I gathered my thoughts quickly and said, "Okay, why don't we both try not to use that word anymore." I was defeated.

Once you start cursing in construction, it opens up a new communication channel with the workers but watch out. It's addicting.

CHAPTER 10

Respect the Trades

I remember once I was doing tenant work on my first project after moving on from architecture. It was 535 Madison Avenue, a thirty-five-story office building on the corner of Fifty-fourth Street, and my responsibility was to get a 10,000-square-foot office space interior done in ten weeks. When I first got into construction, I overcompensated for my lack of experience by trying to be smarter than the workers. That was a mistake.

I met the carpentry foreman on the floor and proceeded to tell him exactly how to lay out the drywall and ceiling, which parts he should build first, how to sequence the carpentry—studs and drywall and ceilings—and thought I had it under control. He nodded and said, "No problem."

I got busy with other tenants in the building and did not come back to the floor for a week. When I came back, very little was done. So, I grabbed the foreman and started to yell at him. "Why didn't you do what I told you to do?" He carefully explained that he did exactly what I told him to do but that my sequence was all wrong, and he had to stop while the electrician and others got out of his way, and so he couldn't finish.

I realized he was right and looked at him in defeat. My floor was now late and it was my fault. Then he invited me out for a drink. He told me, "Don't think you're smarter than everyone else because you went to a fancy school. If you want something done on a particular schedule, tell me when you need it done and leave it to me how to figure it out."

I learned my lesson. Let the trades guide you because they know more about what they do than you do. Because arrogance in construction will kill you.

The trades are very good at what they do, most likely are doing what their fathers and grandfathers did and have pride in what they do. They studied, went to apprentice school, and learned on the job. When they sense you're a snotty Ivy League type, they shut down or—worse—work against you. And it's not right to treat them as less than you. They are super smart at what they do, and what they do is complicated.

Design architects draw fantasy. They have amazing visions and they put them on paper. Then the engineers and working drawing architects have to figure out how to make drawings that tell the contractor how to build it. But they don't know how to build either. They take their best guess at how to draw it. But it's the trades that have to take what they consider to be crappy drawings and make them actually work in the field.

Tradesmen are actually the true artists because they do figure out how to build projects to the vision of the architects. And for the most part the architects don't appreciate it. Let the architects pick up a hammer, a screw gun, a welding torch and try to do this.

I went to an awards ceremony in Washington, DC, years ago. It was the annual Turner Innovation Awards, and Frank Gehry was headlined to receive the award. I had never heard Gehry speak before. To my surprise he brought with him the founder of Gehry Technologies, a company he had set up because no curtain wall contractor could build his projects due to the complexity of his undulating curtain wall designs. Frank Gehry understood the trades, respected the limitations they were telling him that they encountered in his designs, and Gehry decided to help them accomplish his vision. His Gehry Technologies group did the drawings for them so they could fabricate.

Instead of just being arrogant and telling the curtain wall firms to figure it out, he met with them, understood their problem, and solved it. That is what you have to do every day in construction.

CHAPTER 11

Make Sure the Owner Likes You

There is a temptation in construction to blame architects and owners for all your problems. Don't get me wrong, they both create problems. And you are the one who has to figure out the drawings, manage the design meetings, buy out the trades not knowing the full scope of the design, and keep the project on time and on budget. No small feat.

But resist the temptation to blame anyone. As far as the owner goes, usually a very smart developer, you have to learn their management structure, their team, and the decision makers. Usually there is one real "owner," the main decision maker. You have to figure out their personality, take nothing personally, and work at making them like you. Otherwise you can't succeed.

If you don't do that, your life will be miserable. When I built Metropolitan Tower for Harry Macklowe, I learned about managing the owner. Harry was a brilliant but very busy owner. Around him he had a vice president, an architect, and some other assistants. Usually there is the real owner and then the person who executes decisions day to day.

In this case Harry was very involved with design and the early stages of the project and was very good at what he did. He had a lead architect named Bill Derman who was very smart and had to interpret and execute Harry's visions.

Then there was John Tassi, the executive for day-to-day financial decisions. They were all important. I worked very hard to keep them all happy, do what they asked me to do, and complete Metropolitan Tower on time.

There is no secret to working with owners. They are taking a big risk in the project, assembling the land, figuring out if it works financially, hiring the architects and engineers, borrowing the money, and then building the project. All at risk, with no income for years. Whether it's an apartment building, office building, or another project—it is risky for them. Then they hire the builder.

What you don't realize is that for them the building part is the riskiest. They don't know what problems they will run into. Sure, they get a so-called "guaranteed maximum price," but they know that won't hold if there are problems. They worry constantly during construction until the project is finished and making money. You can't be arrogant and think that building it is the hardest part. You have to make them confident that you are in control and know what you are doing. Even if occasionally you don't know what the answer is to a problem.

In fact, when I went to work for developers later in my career, I realized that construction is actually the easiest part compared to all the front-end work in development that I mentioned. You have to be patient, explain clearly what is going on, and always keep them informed and be transparent.

If you find a problem, never keep it to yourself. Get the whole design team and owner in a room and explain the problem. Solve it as a team. Years ago, I had a project that was a ground-up office building in Washington, DC. Seemed easy enough. I was the project manager in charge and it should have been easy. Until we ran into two million dollars in contaminated soil. We found benzene from what must have been a gasoline or oil spill. No one knew. It meant stopping the project for over a month.

Rather than panic—we really should have known from the borings—I gathered the team together, got the right environmental engineers on board, and set up a plan. But believe me, this was a disaster for the schedule and budget.

You have to remember that it's not your fault. There will always be challenges in construction. It comes with the territory. When the owner sees that you care—that you are diligent and professional and concerned about their money and their investment—they will work with you in a positive environment. And don't forget to go drinking with them. Developers are drinkers; it's the only way they can get through the stress.

CHAPTER 12

Put in the Hours

If you want to go into construction, you better be prepared to put in the hours. It will take time away from your family, it will stress you out, you will be exhausted, but if you love what you do, it won't matter. If you aren't willing to put in the time, find something else to do.

Don't underestimate the effect on your relationship with your wife/husband or girlfriend/boyfriend. In construction, work comes first. I know because I did not see my children as much as I should have when they were young.

Building a large complicated project is all-consuming. You don't turn it off when you get home. The emails keep coming, the calls keep coming, and you worry about the next day's and next week's activities. I built several projects for the Trump Organization, the first being Trump Tower when I was thirty and more recently the Trump International Hotel in Washington, DC. I can tell you that they are not unique in being demanding and expecting you to be 24/7 involved with the project. It was not unusual to get a call from Mr. Trump or Ivanka (she ran the project for the family) late at night or check my email before I went to bed to find out I had a problem and would not finish until 2 a.m.

But I never resented it. I actually loved it because the projects were so amazing.

And don't hesitate to come in on the weekend to the project site. There is too much to do on a $200M construction site with thirty architects and engineers and hundreds of documents. There are changes to the architectural and engineering drawings every week; there are Requests for Information (RFI) and shop drawings that are constantly moving in and out of the project. No matter how many assistants you have, you as the lead have to be on top of everything.

Even now at age seventy, I work on the weekends. It gives you an opportunity to walk the site, go to every floor, look at a particular area or issue on-site, and take the time to analyze and solve the problem. You can't do that by email or from home. You can only do that at the project site.

I was senior executive in the early 1990s for Morse Diesel International, a large builder in Manhattan, on a project for Salomon Brothers HQ at 7 World Trade Center. It was one million square feet of interiors in a two-million-square-foot building- a $250M project. It was very intense, twenty-six floors of complicated renovations, creating two-story high trading floors from existing one-floor spaces, high-voltage electrical systems, independent HVAC systems, emergency power generators, executive suites for the big shots, cafeterias, and dining rooms.

I had a staff of forty and the owner had another fifty people on-site. We had thirty architects, engineers, and consulting firms. We had meeting after meeting after meeting during the week. We had two and a half years to build it.

One time I had to come up with a budget for a new subproject in one day. This was no easy task. I stayed in the office all night with no sleep and got it finished by an 8 a.m. meeting I had with my boss. When I met him, he joked, "You look like shit. Did you sleep in the office?"

I told him, "Not exactly. I worked all night." This kind of take-no-prisoners attitude is what made me successful and people notice.

Knowing what is going on with every inch of your project is critical. The only way I could keep up with the Salomon Brothers HQ was to come in on Saturday and walk all twenty-six floors and then catch up with paperwork. In construction there is no substitute for hard work and hands-on management. Today it is too common for project managers and engineers in the office to assume their superintendents are doing the right job and not go into the building.

It is understandable with all the emails, shop drawings, interactions with the architects and engineers and other paperwork that a project manager has to do. But don't be fooled. If you don't work harder than everyone else—be willing to roll up your sleeves and get dirty—and think you can just stay in the office and send emails, you will not have a good career in construction.

CHAPTER 13

Talk to People

In the last chapter we talked about putting in the hours. The other part of that is talking to your project team, foremen for the subcontractors on the project, architects, engineers, consultants, and the owner's staff. You only get to know what is going on in a project by talking to everyone as you walk around, go to meetings, talk on the phone about project issues, and engage. And when you do, people appreciate that you care about the project and want to discuss it every day.

I was the owner's representative recently on a $250M expansion to the iconic Kennedy Center in Washington, DC. This was the first major addition in fifty years. This project was complicated because the Kennedy Center had terminated one general contractor in the middle of the project and brought another one on board. I was brought on at this time as we had to transition and deal with claims from every subcontractor that continued with the new contractor, Whiting-Turner. At the same time, as the project was late and over budget, we had to move forward as seamlessly as possible.

When I came into the project, I knew no one. The owner was very professional and ran a million-square-foot facility with many theaters.

They didn't know me. I was there to get the project on the road to recovery. My résumé looked good, but they didn't know if I would succeed or not.

The only way I have learned to know a project is by talking to everyone, finding out what the priorities are, and planning the work. You don't do that by computer, email, or a software program. Since I have had the good fortune to have been an architect, contractor, and developer, I understand all sides of the building process. Each has a different priority and goals.

The first task was to get to know the contractor, Whiting-Turner. I had only worked with them once before on a very small project. This was a rough-and-tumble team that had just built a billion-dollar casino in Maryland. My first day I met the general superintendent, Buzzy Driscoll, an imposing man with forty-five years with Whiting-Turner.

Talking to him was going to be tricky, as I knew he didn't trust owner's representatives. Our experience with owner's reps in construction is that they don't know how to build, like to blame everyone other than themselves for the problems, and don't move the project forward.

As soon as he met me, he had refused to give me any information and basically ignored me. I looked at him and said, "I don't really give an eff what you think of me. I have a job to do and if you don't like it, I'll do it without you. You're not my effing boss."

As I walked away, he came up to me and said, "Were you a contractor at one point?" I confirmed his suspicion and he said, "We're going to get along fine." From that point on he worked with me.

My point in this is you have to talk differently depending on who it is. The thing I love about the building world is that you can be talking to a Buzzy at one moment, an architect the next, and the Yale-educated owner the next. If you are good at it, you can switch gears in a minute. And if you don't love talking to each of them, you will fail. You have to love what you do. Construction is about relationships.

CHAPTER 14

Get Dirty

In addition to getting along with everyone on every level of the jobsite, you have to be willing to get dirty to be successful. That means when a subcontractor, carpenter, concrete worker, foundation sub, or plumber has a problem, you have to be willing to get into the trenches and look firsthand at the problem.

If you aren't willing to do that, you lose the respect of the trades. A few years after I went into construction, we were having problems with leaks on the outside curtain wall of a forty-story tower in lower Manhattan, 40 Broad Street. The ironworker told me the only way to see the issue was to climb out on the swinging scaffolding from the roof and lower down to the window.

I was not great with heights and as an architect had never had to go out on a scaffold. It was a fairly windy day, but in order to save face I said, "Sure." Had I not agreed the word would have gotten around on-site that I was a "chicken" and was afraid of heights. So off I went in my harness tied off to the sides of the moving scaffold and went down from the roof.

We were examining the problem on the west side of the tower at floor thirty-six, so we stopped there. The ironworker knew I was not crazy about being there and tried to help by saying, "Don't worry, if you fall it will be over really fast." Not helping. Then he added, "Just look straight ahead and pretend you are on the ground." Now that helped some.

During the inspection a gust of wind came up, and we moved away from the building about five feet and then banged back to the building. "That happens all the time, no worries," he said. The mere fact that I didn't throw up or pee in my pants meant I was okay by him. So that was my test and I passed. My hands didn't exactly get dirty, but you know what I mean. If you aren't willing to crawl in a dark space, go on a high scaffold, climb up a ridiculously tall ladder—then you can't succeed in construction.

Another time I was working on a 12-story building in Washington, DC, that had a large footprint of 45,000 square feet. It had a very deep foundation, seventy feet, with a mat foundation—a very deep concrete base that held the building down. The excavation was done, with steel H piles and wood lagging holding the dirt back before the concrete walls were poured.

The only way down was a not very stable wood stair that they had built on one side. On every floor there was a wood platform, but you definitely felt like you could die going down this. You will find yourself facing this all the time. So, I went down to the bottom, looked at the issue with the tieback that was slipping, consulted with the engineer and contractor, and we went back up. Problem solved. If you don't see it yourself, you can't understand what decision to make. Pictures don't do it.

That being said, there is no place for daredevils and bad judgment. If it is really too dangerous, then don't do it. Rarely would a construction worker ask you to do something that is not safe and that he would not do himself.

CHAPTER 15

Have Integrity

Integrity of character is essential in construction. If you promise to do something, make sure you follow through. On construction sites and with the other team members like architects and engineers, you develop a reputation very quickly. Everyone learns whether you are a stand-up person or someone who moves the goal posts for your own benefit. Even if it doesn't work to your benefit, honesty and transparency are important.

If you follow this rule, most people will reciprocate. They will learn that you are a man or woman of your word. And it will make your life much easier when you approach a complicated $200M project.

This also means keeping people's confidence when they tell you not to share information. Transparency does not mean sharing everything with everyone. Especially with the owner, where there is a lot of confidential information. It may be tempting to tell a subcontractor some inside information that they would love to have. But trust me from my having made mistakes over the years, it always comes back to bite you.

And this is especially important with bidding and soliciting pricing from subcontractors. During the initial bidding process, you will have all the pricing for all the trades as you put together the GMP or cost of the work. The subcontractors will be relentless in trying to find out if their bid is low or high. In many cases you will have a long-standing relationship with the sub and be tempted to help them. It may seem harmless at the time, but it will hurt the bidding process in the end. You will develop a reputation as someone who cannot be trusted if you start sharing pricing or budgets. Believe me, in construction everyone knows everything about you.

Early on in my career, I learned that when dealing with building and the trades, if you promise something, you need to keep to your word. I worked with a carpentry foreman, Tony Giacomo, on a high-rise at 750 Lexington Avenue in the late 1980s. It was common for us to do favors for each other. Sometimes I would need him to do some extra work outside of his contract and then settle the money part later.

Trust was everything. If I had to wait for the paperwork for him to redo some framing on a floor where the owner wanted a change, it would delay the schedule. I had to get it done right away. He would always proceed with the change right away because he knew I could be trusted. Had I not followed through later with the proper paperwork and made sure he got paid for the work, that would have been the end of getting things done on time. And all the other foremen would have stopped doing work for me also, because it's a band of brothers and they talk to each other.

The other part of integrity is making sure you don't take cash on the side from the subcontractors. Don't do a favor and then take a "reward" from them. In the old days in New York City, this was common and hard to say no to. In some ways you were disrespectful not to take a "gift," especially from the Italian trades.

On my second project in Manhattan, I worked with a concrete subcontractor. We all knew they were mafia, but we were supposed to pretend they weren't. The mafia was mostly in the trades heavy in labor like concrete, masonry, marble, carpentry, and plaster. And in New York City at the time, they were all union trades. They were very good trades, got the job done better than any contractors I have worked with since then, and nice guys if you didn't cross them. That didn't mean you couldn't argue with them; you just had to be straightforward and not play games or be sneaky with them. That would really not be good. It was not a problem for me as I was a straight shooter from the start.

One day, Christmas in 1983, my superintendent, Charlie, came in and handed me an envelope. We were working on a sixty-five-story mixed-use project on Fifty-seventh Street. I knew what was in the envelope from the thickness of it but told my super I was not interested in taking the "gift." He insisted it was fine because it was Christmas. But it didn't feel right to me. I called my boss at the main office, and he told me, "Take the envelope." Apparently, he was not a good role model.

Now this was not as simple as you think. The Italians in the construction trades took this kind of thing personally. I had just insulted them by refusing the envelope. What to do? Some small amount of panic set in. I slept on it and came up with my strategy—I would turn the Mafia code around on them. As I knew would happen, Anthony from the concrete company came by to see what the problem was. He didn't mention the money until we were at lunch and then said, "What the eff is this about not taking our Christmas gift?"

I had practiced my response. "Anthony, I was insulted that you sent a messenger and didn't give it to me yourself." Now I had turned the tables because he knew I was right.

He said, "Okay, you're right, so here you go," and handed me the envelope. I then acted petulant that I was insulted and wouldn't take it. He knew what I was doing and let it go. Close call.

CHAPTER 16

Be Organized

This is the part of this book where I do recommend referring back to your studies, training, and the many good construction management books. An excellent short book I like is *Successful Construction Project Management* by Paul Netscher. In this easy reading book, he covers everything from procedures, jobsite management and safety, schedule, budgets, quality control, working with owners and subs, and proper paperwork. Written in 2014, it is still quite relevant even as the technological side of construction takes over.

I also like the classic management book by Peter F. Drucker, *The Practice of Management*, written in 1954. It is still good for any construction manager, because organizing and running a complicated $100M or $200M project takes an understanding of management and organizing people.

What they will all tell you is procedure matters; proper organization matters. When I first started in construction, I was an assistant project manager under an old-time project manager. I had just come from three years practicing as an architect. I knew a lot about design and construction documents from the A/E side, but very little about how to actually build a building. My education at Columbia University's Graduate School of Architecture and Planning had only provided me with one course on construction, mostly materials and some technology.

My first project was building out tenant work on a thirty-five-story office building on Madison Avenue. I was given two 30,000-square-foot floors with three tenants. The schedule was tight and the budgets were fixed. I remember panicking because I knew what I didn't know. But this project manager was very organized and had been building for twenty years. The first thing he told me was to stay with procedures. Review the drawings, get pricing from subcontractors, establish the schedule, review with the tenant, and get them to sign off on the schedule and budget.

Set up the plan and follow it. Don't cut corners; don't make things up as you go. Follow the plan. I have remembered this over the forty-five years I have been doing this. Basics matter.

This goes for meetings also. Whether it is a design meeting, owner's meeting, or subcontractor's meeting—set a clear agenda, follow the agenda, hold people accountable in the meeting, stop the meeting from getting off course, and follow up with clean meeting minutes with action items and dates.

This will take discipline and sometimes being pushy. I have a good sense of humor and would often use that to get a meeting back on track. Years ago, I had a superintendent named Frank Bavasso. I loved working with Frank on this 265-unit, ten-acre housing project in East New York (near Queens). He was a natural genius for building but didn't like procedures or being structured. He certainly didn't like a kid twenty years his junior bossing him around. But I was responsible to the company and the owner to deliver the project on time and on budget. Supers tend to worry only about getting the project built, not the money.

In meetings Frank would go off on tangents, telling stories, talking about small details, while I only had an hour or two to get through my agenda.

BE ORGANIZED

Because Frank liked me, I could cut him off nicely and refocus on the agenda. But believe me it wasn't easy. And if you don't get along with the super as a project manager, it is impossible to rein that in. People are well intentioned, but the biggest problem with meetings is going off track.

Recently I was owner's representative for the Kennedy Center's expansion project, The REACH in Washington, DC. We had twenty architects, engineers, consultants, a general contractor, and thirty subcontractors. There were often five or six meetings a week, some run by me and some run by Whiting-Turner, the general contractor. And then there was the weekly owner's meeting and quarterly Kennedy Center Board meetings.

If you don't set up a clear agenda, prepare the visuals and handouts, and work quickly through the topics, you can't build. I believe you either have the skills to manage this kind of organized chaos or you don't. Some people are good at it. Many lose control, and the meetings don't accomplish what they were intended to. And this leads to late projects and wasting money.

The other lesson I learned from Artie Nusbaum, my mentor, is to save every minute in construction you can. He used to say, "At the end of the project, you would kill for an hour or a day to finish when early on you think it doesn't matter." And the key to that is organization and staying with the plan. As a construction manager on a large project, you have to be hovering above in the helicopter—not in the weeds, where you can't see the big picture.

Every day take a look at the schedule and cost. Always see the problems before they arrive. There is nothing worse than getting to the end of the month when you, as a construction manager, have to deliver the updated cost report and schedule to the owner, and you find out that you are late and over budget. You should never have that sneak up on you.

CHAPTER 17

The Dangers of Email

In case the younger construction managers and engineers think I'm too old school and out of touch with technology, be assured that I am very conversant with cost control software, Excel, Word, Primavera Project Management, Procor, CAD, Bluebeam, PlanGrid, etc. I even text, much to the surprise of my daughters. And all of these, including email, are good tools if used properly.

That being said, they are not a substitute for rolling up your sleeves and getting involved, calling subcontractors, architects, engineers, and consultants and walking out of your nice warm office and seeing the issue firsthand.

I can't tell you how many times I have heard a response from younger staff when I would ask the status of an issue that they "had emailed them." As if that solved the problem. In the early 2000s, I was president of the Mid-Atlantic office of Tishman Construction. I had a staff of forty smart, mostly young engineers. We were running ten projects in the Washington, DC, area. I had to constantly be aware of the status of each project, as I interacted with the clients and was always chasing new work, usually from existing clients. We could not afford having a problem on a project that we were not ahead of or solving.

I would get copied on almost everything and would, with the help of my administrative assistant, track any issues that I thought I could help with or would cause a relationship problem with an owner. So, I would either walk around the office and ask questions, call a site and talk to the project manager or superintendent, or contact the architect, engineer, or consultant myself to find out what was happening.

More times than I can remember, the answer I would get from my young staff would be the standard "I emailed them." I would then say, "Is the problem solved? What happened?" And too often they would say, "Well, no, I am waiting for them to email me back." Meanwhile the clock was ticking on the schedule, some trade was being delayed, or it was costing money by not resolving the issue.

The other thing that would drive me crazy is one of my staff coming to me with a problem having given no thought to the solution. I learned this early on when I worked as an assistant project manager at Trump Tower in Manhattan in the early 1980s. My boss, Aldo Rizzo, had a temper and was short on patience. The first time I brought him a problem without having options for the solutions was my last. Aldo was right—he had more than enough problems building this seventy-story mixed-use project with Donald Trump as his client without me adding to them. So, I learned to have as many options as possible if I went to him—or solve the problem myself.

Sometimes when you are new to a project or construction in general, you are afraid to make a mistake. We have an expression in construction: "If you're going to fall, fall forward." It's often better to make a decision or solve a problem quickly, or the problem will take its own course and be decided without you.

Your job as a construction manager is to solve problems quickly. Time is your enemy. Pick up the phone, set up a meeting, visit the architect or engineer, bring in the consultant. Do your homework and have options before you call. Put everyone you need in the room and fix the problem. Emailing is not the answer.

CHAPTER 18

Don't Be Pushed Around

This may seem obvious, but you would be surprised how many graduates of construction management, engineering, and architecture programs are too nice. Superintendents who come out of the trades don't seem to have this problem as they have learned in the School of Hard Knocks how to succeed.

I have worked with many young project managers who think by being kind and polite they can get things done. I'm not saying you have to be an awful and mean person all the time as some told me in 1981 when I started in construction. Back then I was too nice and had to learn how to be forceful. But I was never mean for the sake of being mean.

I have alluded to this in past chapters, but with the trades, subcontractors' project managers, in their main office and even with owners, they will see blood in the water if you can't fight when it is appropriate and back off when it is better to do that. It is a balancing act that, again, the "construction personality" has an instinct for.

I have been in screaming matches with foremen of trades like electrical and concrete that an outsider would think were the end of the world. They usually think they are right and I, of course, think I am. Yelling in construction is how we communicate. There is a lot of pressure and the stakes are high. The foremen have to watch their labor, make sure materials arrive on time, and they are constantly being scrutinized by the owners of their companies. They look at production rates every day. The profit margins are thin in the tough competition and bidding for work.

That's why pushing hard is a good thing. Not just for your ego and building faster than others. It is good for the trades, who often don't see the big picture—that they are delaying other trades and that you will do the same for them if others are delaying them. And the faster you make them build, the more money their bosses make and the better they look.

A case in point was on 7 World Trade Center, the Salomon Brothers HQ buildout of one million square feet. I was constantly in screaming matches for more labor and better sequencing with a thousand workers on twenty-six floors of construction. Morse Diesel International was the construction manager overseeing two other general contractors, Herbert Construction and AJ Construction as well as constructing $100M of infrastructure ourselves. Herbert and AJ in turn managed forty trades between them for a total budget of $250M ($700M in today's dollars). I would try first to solve the problem with Herbert and AJ but had developed direct relationships with the subcontractors. If I was not happy with the progress on a given day, I would go directly to the foreman at 6 a.m., the best time to find them in their shanties, and have my fight. They had their own issues, because in 1991, things were busy in Manhattan and labor was tight.

Salomon Brothers HQ was a union project as everything was in Manhattan at that time, and the unions controlled the labor. The firms with the biggest workload and political influence at the unions got the best labor. So often these foremen were screaming at their bosses for more labor, and it was not available.

They understood my pushing them hard, but on the other hand it annoyed them. But in construction more than other businesses, the squeaky wheel gets the grease. If you don't fight every day, you will be passed by and walked over. If you are too nice you should, as Artie Nusbaum used to say, "try hairstyling. It's a lot easier."

CHAPTER 19

Support Your Team

When you run a large project or are head of a construction office, you need your team to be behind you. No one can do it alone. You need to have the instinct to know who is capable of performing and who you need to fire or lower the expectations for workload and accuracy.

I was president of Tishman Construction for the Washington, DC, office in the early 2000s. I had just finished a successful run with a developer in Manhattan, advising him on a sixty-five-story apartment building in the Wall Street area. It was the first high-rise after 911 in this area. I was finished and wanted to come back to DC.

I was looking for a new role and ran into my friend Jay Badame in Penn Station in New York while commuting to DC. I had worked for Tishman once in the 1980s and had worked with Jay when he and I were both PMs. Now he was the chief operating officer of Tishman. He told me that they had an opening to run the DC office. I applied later that week, interviewed with Dan Tishman, and was hired.

The DC office of Tishman had forty people running many projects, small and large. The tough part of this assignment was that the number two in the office had hoped to get my job but was hopelessly incompetent and bad with people. So, Dan Tishman knew they needed someone who could sell to clients, run the office, and not alienate the team in DC.

When I arrived, I found out that most of the office couldn't stand the number two. Having learned early on that success is all about the team, I went and met individually with all forty. The office had only been successful before because the previous head of the office was competent and good with people. With the prospect that the number two would get the head job, most of the staff was ready to revolt. The team would surely have failed with him in charge.

None of the staff knew me, but by my talking to each one and letting them know I was on their team and would back them up, I quickly developed the trust to succeed during my time there. And of course, you have to follow up with that trust. The number two would promise one thing and then help his favorites and screw everyone else. I talked to the lead PMs almost every day and kept in touch with the clients to make sure they were happy.

Another time I was at a smaller firm, York Hunter, where Artie had become a partner in the early 1990s. Artie had left HRH Construction for a smaller venue. I was project manager working on several construction projects and consulting all over the country with Artie. Here again I took the time to get to know everyone, to work at having them trust me and socialize regularly with all of them (see Drink with Your Team, Chapter 8).

That means from the receptionist, to the administrative assistants, to the superintendents in the field, to the project managers and accounting staff in the office, they all are critical to your success. Try building a project fast if you don't have a great relationship with the accountant and CFO in the main office. Getting checks out quickly to subcontractors, cutting special checks for materials out of sequence, and generally making sure the accounting is perfect are as important as the construction on-site. The team runs from the main office to the project team on-site to the subcontractors building it. They are all part of the team that will allow you to succeed or fail.

CHAPTER 20

Say Yes to Travel

My aunt, Margaret Dalton, never got married, and her nieces and nephews were really her children. She always gave good advice. As the assistant to a senior executive with Western Union, she saw many candidates for jobs come through her doors. And one thing she always told them—and she told me—was "Never say in an interview that you don't want to travel. It's a deal killer." Many a candidate did not get a job there because they said they didn't want to travel. They might never have been asked to travel or relocate, but by saying they wouldn't travel, they gave another candidate who said yes the edge.

Travel, whether it is to check on a factory that is making critical material for your project or visiting an engineer or architect in another city or country, is part of your role as a construction manager. And sometimes relocating is necessary to avoid being stuck in a rut in your firm.

In 1992, I was working for Morse Diesel International having just finished the Salomon Brothers HQ project on time and very successfully. I was in the main office without another project on the horizon. When my boss came to me and said, "We need you to travel to other offices. Is that okay" I said yes right away. At the time I had a wife who was not working and two young girls at home. I was not about to lose my job by saying no, particularly with no project at the moment.

That even involved moving to Hungary for two months to assist a Hungarian construction firm with bidding a $50M office building for Olympia and York, a huge developer from Canada founded by Hungarians. Although I didn't want to be away from my children for that long, I showed that I was flexible and loyal to help the firm out. And they knew I could handle the support that we needed to provide to this construction firm in Hungary who was not used to bidding competitively. It was just coming out of a Communist business structure and had never competed for a project.

It was a great experience, I learned a lot, and we won the project. After two months in the middle of Hungary, I returned home to New York. It was at that time that I started assisting the DC office of Morse Diesel International. I would fly down for a week and then come back to New York. Again, travel allowed me to show the firm I was loyal and flexible. Very important when you are running a large office.

While in DC I learned that the head of the office was going to move back to Germany and take another job. That information and the willingness to relocate allowed me to advance in my career. After a few months working in DC, armed with the knowledge that no one else in the firm knew that there would be an opening to head the DC office, I had a plan.

One day I made an appointment with the chairman of the board, Don Piser, to discuss the DC office. Mr. Piser did not even know who I was. Because I always made friends with the administrative assistants as well as everyone else in the office, I got an appointment right away.

When I met with Mr. Piser, I was very nervous but laid out how I would be the best person to take the job as head of the DC office. He listened carefully, asked me a few questions, and then I left having no idea how I had done. Mostly I thought I had made a fool of myself being too junior to run an office.

Much to my surprise, the next day I got a call from Mr. Piser personally to tell me he was naming me head of the office in DC. He then said, "Do you know why I chose you?" I said no and he said I was the only one who approached him, was willing to relocate without hesitation, and that my aggressive spirit had led him to believe I could run the office. It was a game changer for me and taught me that you have nothing to lose by pushing for yourself no matter what the change may be.

CHAPTER 21

Respect Your Mentors

No one succeeds without others helping you. No matter how lucky or talented you are, without others helping you up, you will stay on the bottom forever. Part of this lesson is that when you are a mentor to someone else (which you should strive to be), it will always come back to help you. And you don't do it for that reason; it's just karma and who you are.

As you advance in your career with the help of your mentors, don't forget them after you move on. Always stay in touch with them and thank them for what they did to help you. I always did this, particularly with Artie Nusbaum, who was my first mentor, up to his death at ninety-two a few years ago.

When Artie turned eighty in 2005, many years after I had last worked with him at York Hunter, my friend John Leeper and I threw a party for him. We had both started with Artie and invited everyone who had benefitted from Artie's guidance and mentorship over the years. Every one of them showed up. We rented a floor at an Italian restaurant on Fifty-sixth Street in New York City. I bought a large coffee table book about skyscrapers, and every one of Artie's mentees signed it on the way into the party. Artie was touched, which for a tough old-time construction guy from New York was amazing, and we all took turns telling him what he meant to us.

John and I didn't do it for our own egos. We did it because we both knew that without our mentors, we would be teaching high school math in suburbia, not building amazing skyscrapers in Manhattan.

In 2015 when Artie turned ninety, I wrote him the letter below so that he would know what he meant to me as a mentor.

September 12, 2015

Artie,

For your 90th birthday I thought I would share with you some thoughts on what you have meant to me personally and all of us who had the privilege of working under you and learning so much from you.

You mentioned to me yesterday about how Carl Morse was so important to you in your career. It got me to thinking that, just as Carl Morse trained a generation of the finest builders in the country, you can take credit for training the next generation of builders. People like Nino Noto, Cary Spiegel, Tony Mannion, Joel Silverman, Kenny Brown, Steve Guzzardi and on and on.

When we met you and worked for you, we were instilled with your passion for the art of building, for perfection, for not taking no for an answer when we were told it could not be done or a sub told us they would not give us enough men for the job. You taught us that failure was truly not an option.

You taught us to critique ourselves as you critiqued us. You taught us to wake up in the morning and love our job as you loved yours. You taught by example; you taught as a tough instructor who expected us to work hard and succeed. You taught us to have pride in our work and ourselves. That is a rare gift.

The mentor never gets thanked enough. But we know that you know we love you and love how you made us into better builders and better people. Whenever I talk to any of my New York construction friends, we always tell Artie stories. Some are funny, some are scary, but they all have one theme—we listened to you, we laughed with you and (sorry) sometimes at you. But these stories are our history. They are the great moments when we solved a problem that seemed insurmountable. When we built faster than our competition, when we built faster than we thought we could.

My favorite story is one I tell often. One time you came to Metropolitan Tower and as usual we walked from the 65th floor to the bottom. It was the usual entourage of supers and project managers. You had asked one of the assistant superintendents to do something in a particular way with a sub for three weeks. He thought he knew better and ignored you. He did not realize that you never, ever forgot what Artie told you in previous site visits. The rest of us knew we either did what you told us because it was the right solution or we challenged you with our best foot forward, but we never ignored you.

So, after seeing that he had ignored you for the third time, you turned to him and said, "It's not too late to become a hairstylist." He was speechless and clueless. After you left, he asked me what that meant. I said to him, "Find another job." He just looked at me and turned his head like a dog that didn't understand. But we understood.

You expected the best from us, you encouraged us to be the best, and when we did hit that homerun or we did solve a really tough problem, occasionally one even you didn't see, you looked at us with pride that your kids, your students had taken the mantle and could now lead on their own. You weren't one of those bosses that was jealous or threatened when your protégés became great builders. You rejoiced in it.

So, on your 90th birthday you need to know that we didn't know what we were getting into when we graduated from good schools and stumbled into your office for the first interview. We didn't know that this would change our lives forever and make us look back and feel that we too are the luckiest people on earth to be in a world where we can take thousands of pages of drawings and specs, mostly done badly, and turn this into urban sculpture. To walk by these buildings years later and say, "We did that. We built this amazing skyscraper."

And finally, I have one more story. My daughter Abigail was born while I was at Metropolitan Tower. My second daughter, Cornelia, was born when I worked with you at York Hunter. And coincidentally you and Marilyn became friends with my in-laws, so my children grew up with Artie in my stories.

One day a few years after we finished Trump Tower, I was walking with Abby and her friend when she was about six. We were passing Trump Tower and she turned to her friend as they both looked way up at this sixty-five-story building towering into the sky and she said, "My father and Artie built that." I was about to explain to her that there were hundreds of people involved but then decided to just say, "Yes, we did."

Happy Birthday.
Steve Dalton

Artie called me shortly after reading this and said something I thought I would never hear from him. Remember, this is the guy who told me he didn't like architects in my first interview. He said, "It's not the many projects that I have built all over the skyline of Manhattan that matter; it is the people I have met and worked with over the years." Never forget that you didn't do this alone.

CHAPTER 22

Dealing with Idiots

One of the bigger challenges is working with team members who are either not very smart or so stubborn in their ways that they block you from getting your job done. I call this, not so nicely, Dealing with Idiots. Not politically correct, but after all, this is construction, not a course in construction at a university.

We all have experienced it, and it is frustrating. I go back to the quote from Artie Nusbaum when I complained about an engineer having thirty years' experience and not being very good at what he did. The answer from Artie was "No, he actually has one year's experience repeated twenty-nine times."

You have to manage team members just like everyone else. If you can get them off the team—whether it is the architect, mechanical engineer, structural engineer, or consultant—and replaced, that is the best. But it's not always that easy. Years ago, I worked with an electrical engineer who was also a partner at a large MEP engineering firm. There was no way I could get him off the project. He was terrible, always blaming others, never taking responsibility himself, and working so slowly I wanted to shoot myself in the head.

I could not remove him, but I did find a way. Since he worked for the architect and they knew he was weak, I made it their problem. I told them I would no longer copy him on anything, would not send him any shop drawings for review, and would only send things to his assistant, who was excellent. After some pushback, the architect agreed to help me in bypassing this guy. It was not a perfect solution, but for the most part worked.

When the idiot is your client, that is even worse. I have had some situations where the team member who works for the owner is a pompous ass and not helpful. In this case you have to try a little charm and maybe some sucking up to get them to understand what you want them to do. Maybe have drinks even though you hate them. The goal is always making the project finish on time. I have not always succeeded in that endeavor.

You can also try to mentor the idiots if their rank is below you. I have many times been sent an assistant project manager or superintendent who was not very good at what they did. With Artie, that rarely happened as he was a good judge of character. But in a company of 500 or 1,000 staff, there will be some weak links. In this case you just have to put them in a small box (figuratively) and limit their ability to screw up the project. This means a larger workload for the rest of the team, but you can't have them interacting on something like curtain wall, MEP, or structure. Keep them off the critical path with something like carpentry, masonry, or interiors where you can watch and catch errors.

CHAPTER 23

Don't Forget Your Family

A career in construction on large, complex projects is all-consuming. I wish I could tell you that I was successful by cleverly balancing work and the rest of my life. I would be lying. The only way I know to be successful is to so immerse yourself in your job that you think about it 24/7. You would be surprised how many problems you solve while not at the jobsite—things that percolate in your brain and solve themselves at odd times. Your family or friends or girl/boyfriends will suffer.

That being said, if you pick the right partner, maybe someone who also has a busy career and gets your obsession, then you may have a chance. If you have children, that is harder. I can't tell you the number of evenings or weekends that I was not there for my two daughters. The soccer games missed and other school events.

One way I tried to compensate was to bring my kids to work when I could. This might have been "Bring Your Daughter to Work Day" or maybe on a weekend taking them to a site (safely). One year I took my ten-year-old daughter Cornelia with me on the weekly flight from New York to Washington, DC, and then to the jobsite. She was thrilled to fly with me, stay in a hotel, and go to the jobsite meetings and walk around.

And it is amazing how children view life. We visited the site with a seventy-foot-deep excavation with a hand-built wood stair going down to the bottom. She asked me if I went to the bottom and I proudly said yes. She then asked if she could climb down to the bottom with me. Of course, the answer to that was no, but she at least felt a part of my life and saw what I did every day. It made a difference when I had to work late or on weekends.

Another time I brought my other daughter Abigail to a ten-acre site in East New York where I was building 250 units of low-income housing for a private developer. The site was large with three concrete platforms on top of which were three-story townhouse units with apartments. She was thrilled as we looked at the excavation equipment, concrete trucks, roads being built, and pipe being installed underground. It was a way to show her what her dad did.

CHAPTER 24

Visit the Factory

One of the first lessons I learned as a project manager was not to trust vendors and manufacturers who are fabricating your materials. Some lie shamelessly, and some are well intentioned on delivery times but get behind and don't tell you.

My first major role as project manager at Metropolitan Tower in New York City was a sixty-five-story high-rise that had a complicated curtain wall system. The building was a rectangle on the bottom for twenty stories and then a triangle shape on top for the final forty stories. It was the first four-sided, structurally glazed curtain wall in New York City. This meant it had no mechanical fasteners—only structural adhesive attaching the glass to the aluminum frame.

This had not been done this way before on a major high rise, and our curtain wall consultant was actually worried about the quality control and performance over time and recommended not using this system. But the manufacturer, Glassalum Engineering out of Miami, had tested it and proved its performance. So, the owner opted to proceed with fabrication.

Artie Nusbaum met me one morning at about seven as he was prone to do. We had just started fabricating the frames for the curtain wall in

VISIT THE FACTORY

Miami, and Glassalum had a large invoice in for payment for stored materials and some fabrication. Artie handed me a ticket to Miami for that day and said, "You are going unannounced to see if they are telling the truth." I asked how that would go over with the subcontractor, whom I knew quite well at that time, and he said, "I don't care. They need to know that we are watching them. Learn this lesson—they all lie."

I hopped on the noon flight out of LaGuardia airport and headed for Miami with no notice. I grabbed a taxi at the busy Miami International Airport and was at the Glassalum plant forty-five minutes later. Glassalum had a very smart and successful partner in New York, Elliot Krackow, who handled the installation and the unions. But the factory was run by Cubans in Miami, and I didn't know them as well.

I walked into the front office, which was in one of the three large steel factory buildings on a large site outside of Miami, and asked for Renato Mazarantano, the partner in Miami. His head of the plant and also a partner was Ralph Reyes, who looked like a Cuban Sumo wrestler and was very imposing. Renato came in, surprised that I was there, and said that Ralph would never let me in the plant. Being the stubborn Irishman that I was at age thirty, I said I really didn't give a damn and proceeded to enter the plant.

Ralph came running after me, stood in front of me, and told me if I took one more step, he would throw me into the paint booth and kill me. That was a bit of a problem. Renato had caught up with us by then and tried to calm the situation. But Ralph was not budging. I told Ralph that we were not paying his $5M pay application because he refused to let me see the stored materials. Then I walked out and proceeded out the front door. Renato came after me, knowing that if they didn't get their payment, it would be a disaster, and got Elliot on the phone.

After about twenty minutes of strategizing, we came to an agreement to let me walk through the factory for thirty minutes and take pictures. If I hadn't threatened to hold their payment, I would never have gotten that far. But I was still mad at Artie for making me go there with no notice.

What it did was let Glassalum know that they had to be honest with us and that we were watching. Most fabricators, whether they are oversees or in the US or Canada, need supervision and to be held accountable. There is no substitute for visiting the factory.

Another time my boss sent me to Toronto to check on the fabrication of a window-washing rig. Again, I got on the plane, headed to Toronto, and arrived at the Swing Stage LLC factory. This time I was expected and met with the engineers to go over the shop drawings and then on to the factory floor. In this case they were perfectly organized and all went smoothly. Usually that's the case.

My favorite story of visiting the factory was Italy. It was one of my first trips to the quarries and fabrication plants in Carrara, Italy. I was traveling with the architect in December to review samples of the marble that would be used for the lobby of a 350,000-square-foot office building in Washington, DC. We flew overnight and arrived at the plant in Forte dei Marmi in Carrara about noon. On this trip I learned to love the Italians but also to not trust all of them. As soon as we got there, the owner of the marble fabrication said it was time for lunch. I tried to insist we look at the samples, but this is Italy and you don't get in the way of lunch. So off we went to one of his cousins' restaurants and proceeded to have a five-course lunch, or what the Italians just call "lunch."

By now it was 4 p.m., we were three bottles of wine in, and the sun was setting—but it was time for an espresso. The other lesson I learned was not to get in the way of coffee with Italians. When we finally got to the samples, large slabs of marble about eight feet long and four feet high, it was totally dark, and much to my surprise, they were outside with no lights. The owner said no worries, and he proceeded to bring two trucks over with their headlights on.

Now I knew as an architect having reviewed many samples that we were screwed because the headlights had a yellow tint, and we always wanted sunlight for true color correctness. At this point I was drunk enough that I gave up, the architect approved the samples in the dark, and we went to the hotel to check in. By the way, three hours later we were at a wonderful restaurant outside of Forte dei Marmi and on to another four-course meal. Luckily when the materials arrived in the US, they were perfect and all was good.

Another way to control the quality and timeliness of factory production is to hire a local inspector to monitor the work. This is a cheap and effective way, particularly when you are producing overseas. I have done this with marble in Carrara and millwork in Germany. The only thing you have to watch out for is that you don't hire the brother-in-law or cousin of the owner of the factory without knowing it. This has happened to me in Italy with really bad results. And if you don't control the quality and production rate, by the time it gets to the US, it is too late to reject it and start over. It is money well spent if you find the right person.

CHAPTER 25

Promote Yourself

I always interpreted the expression "God helps those who help themselves" as meaning no one will promote you if you don't do it. I think religious leaders would disagree and think that's selfish, that it should apply to more altruistic endeavors, but in construction as in any field it's true. This goes hand in hand with having a mentor to help you up. But if they don't see you, then you may never find a mentor. This has been true for my career.

When you first start out with a large construction firm, you are a nobody. HRH Construction and later firms like Tishman Construction and AMEC in my career have hundreds and sometimes thousands of employees. My experience interviewing with Artie showed that when he said, "We don't like to hire architects, but we will give you a chance." After a year with the firm, building tenant work and doing what I thought was a great job, I was invisible to the main office.

My boss on my construction site was not particularly interested in helping me or letting the main office or Artie know I was performing well. That will happen too; your boss will selfishly take credit for everything that goes well and blame others, meaning you, for mistakes. So, after one year I got up the courage to get an appointment with

Artie, the president of HRH Construction. Artie was good that way. His door was always open and he would always accommodate you if you insisted. Being friends with his assistant, which I was, helped too.

So, in I went to his office and made my pitch for a raise and more responsibility. Artie listened and then looked at me and said, "I'll be honest, you're just another warm body around here. Go back and do your job and be patient." Well, that didn't go as planned. But I was not discouraged. I went back to my jobsite and worked harder.

Persistence is important. A year later I was on a tenant project in Wall Street for HRH and had singlehandedly run 250,000 square feet of tenant work with really bad superintendents and no real help. But Artie watched everything and knew what I had done. Again, I made an appointment with Artie, marched into his office, and asked to become a full project manager running a base building. I had been pigeonholed into tenant work because I was an architect, and the firm thought only engineers could build from the ground up. Then came another great Artie quote: "You're not ready yet to be a full project manager; you don't have enough experience." I left feeling discouraged this time but did not give up.

As it would turn out, two weeks later the project manager for Metropolitan Tower on Fifty-seventh Street had gotten into a fight with the owner and been thrown off the job. Artie called me up and asked me if I wanted the job as project manager for this sixty-five-story, ground-up, mixed-use building. I couldn't resist and said, "It's amazing how much you can learn in two weeks. I guess I'm ready now." Artie gave me the usual string of expletives, and I hung up the phone happy that I had achieved my goal.

Had I just waited for the opportunity, it would not have come along. It was only my persistence that paid off—getting in front of your boss or upper management and taking a chance. Put your ego away, suck it up, and try. And don't give up.

CHAPTER 26

Don't Burn Bridges; Solve the Problem

Construction is truly a field where karma plays a part. It's such a small world that you can build a project in New York and then years later build one in Los Angeles and you will run into someone you know. This happens to me all the time. I think this is good advice in any field, and many people don't follow it: Don't burn your bridges.

This goes for people who work in your construction firm above you, at your level, and below you. It goes for the trades and the architects and the engineers. It goes for the consultants who help you succeed on any project. It will be tempting to burn people along the way, I know (see Chapter 22: Dealing With Idiots), especially if they get in the way of your progress. But if you are going to do it, be careful and don't do it very often.

I am an arbitrator with the American Arbitration Association. Very few architects are approved as arbitrators, but I was fortunate enough to be one of them. The training required to get certified and stay certified is excellent for learning about coming to solutions with parties who often are at odds with each other or even hate each other.

You will come to a situation where the subcontractor is dug in and will not move from a position on money or documents or schedule. Or maybe it's a dispute between the owner and the general contractor. It would be easy to burn bridges here and walk away with no solution. But you have to remember your goal is to build faster and smarter than everyone else while staying within the budget and schedule.

In arbitration each side takes turns presenting their case, maybe with a witness or two, then gets to rebut the other side. And then the arbitrator, as judge and jury, decides on the merits of the contract and facts.

This served me well and was what I was used to doing anyway. You have to put everyone in the room or on the video conference call, have each side present, and then have a discussion with ideas that make everyone reasonably happy. When a subcontractor is forced to do something that they are unhappy with, it does not go well. They will find another way to punish you later, as they are smart and can read the contract also. Better to give a little with everyone.

In most cases you can put a difficult architect, engineer or subcontractor on a call or in the room with the general contractor and other consultants and use creative thinking to come out with a solution everyone buys into. In one case we had a very difficult problem with the doors and hardware, as we had a financially shaky subcontractor in over their head and a design architect who was reluctant to compromise.

We had a very talented project manager for Whiting-Turner, Hannah Marcin, who had the patience of a saint and did her homework. She never came into a meeting without knowing every detail and was ready with several solutions. Preparation is everything with solving complex problems. And hard work and late hours are the only way you succeed. Hannah had that special instinct to never give up until the problem was solved and the work was back on schedule. This was immediately clear upon meeting her.

When I ran Tishman's office in DC, I hired and fired and would go through hundreds of résumés and dozens of interviews. It was at that time that I understood how Artie could size me up in a minute and decide I was right for construction.

CHAPTER 27

Think Vertically and Horizontally

What do I mean by this? I'm not talking about a building structure, although that would be true also. I'm talking about planning. While you are planning the day-to-day events in the weeds, you need to always step back and look at your building project from the helicopter view. Too many times we are so caught up in day-to-day emergencies that we lose track of the big picture.

One way to do this is with the Monthly Report that I would write for the construction firms I worked for. This is an opportunity to update the construction schedule and cost report with everything that happened during the past month. It requires staying up to date with your cost logs, change orders, and anticipated costs and looking in great detail at the CPM schedule. It is also a good opportunity to document what happened the past month.

On most projects I had an accountant who logged every pending and executed change order to the owner. They also logged every submitted change order from the subcontractors. Don't fall into the trap of deciding that a subcontractor's request for a change in cost or schedule is not valid, so you don't enter it into the log. Many times, what you think is not a legitimate cost turns out to be right because you didn't have the right information or maybe your staff did not tell you the real story. Enter it, don't let it sit too long before analyzing it, and then you can discount it, void it, or turn it into a real change order.

There are many accounting programs to accomplish this, usually a proprietary system in your firm. But as they say, garbage in, garbage out. If you don't enter these changes intelligently, your monthly cost report will not tell the real story. The worst thing you can do is lead an owner or your main office to believe that the project is on target with the budget and schedule and then drop many changes in one month that were pending or lost. This will not go well for you.

Transparency in cost and schedule is critical. It may be bad news, but the longer you put it off, the worse it will get. And in your monthly report you will have an opportunity to discuss why costs changed. It usually is not from any error by you or your team. Many times, I have told an owner, "We are just the messengers. Don't shoot us for telling you what you did to add cost and time to the project or some issue that was out of everyone's control." You have an obligation to tell the real story, no matter how bad it may be. You need to train your accountant to do this and make sure you review in great detail every week the cost reports on the server or a printout.

Your accountant is not a builder, and it may not seem like he or she has the same level of importance as the superintendents and project managers who are in the trenches every day, but believe me, a bad accountant will haunt you for the entire project. And a good accountant will save your butt every day. They are looking at the cost and payments from a more objective vantage point and will alert you to things you missed.

Scheduling is the same thing. On most large projects I have had a scheduling expert who is on-site and updates the schedule in a transparent way. They don't care if you are early or late on the schedule;

they are just attending meetings, looking at the site in walk-throughs, and entering data on the critical path. And when it moves the schedule, they will tell you why. Believe me, when you are in the weeds, you will miss the fact that the schedule is in jeopardy. It may be a fabrication delay, a field issue, one sub delaying another sub, an accident, or just something taking more time than anticipated.

Without this regular update telling you where you are in time, you'll find out too late that you are in trouble. The CPM is a great tool and most project managers can't create or enter proper data on Primavera Project Planner or a similar scheduling program to keep track accurately. The schedulers, either in house or a consultant, train for many years to do this right. Often, they are not builders and need your staff's input to accurately create the schedule. Many times, the schedule is put together by a grumpy old-time superintendent who does not care to ever learn how to operate Primavera. But the super, sitting for hours and days with the scheduler, will create an accurate road map that will allow you to track the schedule every week and every month.

On the Kennedy Center REACH project, Whiting-Turner was the general contractor. They were smart enough to outsource their scheduling to a firm led by Tony McHale. Tony had been a builder and then decided to set up his own scheduling firm. The REACH had 4,000 activities and was a very complicated arts center with a white concrete sculptural structure and complicated architectural details. No two walls met at ninety degrees, and the building was constructed on a large site. Without Tony setting this up properly from the start, Whiting-Turner never could have known where they were on the schedule month to month.

The superintendent, Buzzy Driscoll, didn't know how to operate Primavera, but he damn well knew how to build. It was in his head, instinct developed over forty-five years of building. The team of Buzzy and Tony created an accurate road map to deliver the project on schedule.

Now for the horizontal part. It's great to think big, looking at the view from the helicopter. But without being in the weeds day to day, it will be misinformed and paint the wrong picture. Remember—garbage in, garbage out.

Every superintendent and every project manager has to have their trades and territory to follow every hour of every day. The best way to set up a project—we know from decades of experience—is to assign areas to superintendents and trades to project managers. Sometimes on larger high-rise projects you may assign trades to superintendents, but it depends on the project.

And to stay horizontal in thinking, to keep up with every detail, you need to walk your area every day as a super. You need to review your submittals and open issues if you are a project manager. You need to stay on top of your logs and push the architects and engineers to move paper and solve problems that come up with requests for information or publishing new sketches or drawings as needed.

The best project managers and superintendents can tell you any time you ask them exactly what is happening with every inch of their territory. If they can't it's a serious alarm bell (see Dealing with Idiots). You need to train them or remove them from the project. Artie Nusbaum used to walk with us every week from the top of our seventy-story buildings to the bottom. He would ask probing questions as we moved through the building.

He could simultaneously keep the big picture in his head and look at minute details. One of his tricks was to say to one of us, "Why is this trade behind schedule?" or "Where is this material?" If we said to him, "I've been calling the subcontractor five times a day," Artie would say, "What's the phone number of the subcontractor and who are you talking to?" If you didn't know the number off the top of your head, then he knew you were lying. If you truly called five times a day, you would know the number. There was no substitute for being in the weeds, thinking horizontally, and pushing the subs.

This is where you either can think vertically and horizontally at the same time or you can't. I really don't think it can be taught; you either have that instinct, where you will not allow yourself to fail keeping track of everything, or you don't. And if you don't, try hairstyling.

CHAPTER 28

Persistence

I really liked the movie *The Founder*, about Ray Kroc, who founded McDonald's. In one scene in a hotel room by himself, when he was an unsuccessful salesman selling kitchen equipment in his fifties, he notes in a voiceover that the key to success in business is not being smart or lucky. It is persistence. You will get knocked down, and you will fail many times—but you keep going. You keep the vision in your head and never give up. You change the rules as you go. In one scene later in the film, a real estate businessman meets Ray at a bank while he is trying to borrow money before McDonalds had taken off. He catches up with Ray and tells him that he's not in the fast-food business; he's in the real estate business. Ray needs to buy the land and properties for the restaurants so that he has total control and will make much more money. Ray had not thought of it but knew he was right. That man became a partner and a millionaire.

This is true in construction. Back to Chapter 1 where I said, "Be afraid," I explained that you need to always look for the problems before they come up, and when the ones that you didn't anticipate do occur, you push your way through them, think of new ways to solve them. Again, I think of Artie Nusbaum, who taught me that there is always an answer to a problem. You may spend hours trying to resolve it and think you are stumped. We used to say you could put Artie in a room with no doors or windows, leave him alone for an hour, and you would find him on the other side.

I learned from Artie that you need to throw out your assumptions of the way out of a problem. A case in point: I was building this project in East New York which had a 50,000-square-foot concrete platform with stick-built townhouses on top of the deck. The area underneath was for parking. We had to finish the project in eleven months so that the owner could get very valuable tax credits and transferrable air rights that could be sold for high-rise buildings in Manhattan. Mayor Koch figured out he could get private developers to build low-income housing at no cost to the city.

We were never going to finish on time and I saw no way out, because the deck had to go first and it was taking too long. One day my super and I were looking at the problem on site and we realized we could build the deck from each end at the same time with twice the crew. It would save us weeks.

We thought about all the reasons why it wouldn't work: Not enough manpower. What about getting the concrete from the plant? What about the utilities underneath the deck that had to go first? But we kept at it, figured it out and finished on time.

We box ourselves in with preconceived notions of how we think we will build the project or solve a problem. Every once in a while, you need to turn those assumptions upside down, usually with your smart team members. As I advanced in construction and operated more in an executive role, I would often go to meetings and throw out what seemed to everyone there as crazy ideas. I did it intentionally to get them thinking outside their comfort zone. We invent new ways to build all the time this way.

CHAPTER 29

Managing the Trades

There are two parts to managing the trades: managing the owners and main office staff of the subcontractors and managing the foreman and his workers on-site. This is another area that can't be taught. How to juggle thirty subcontractors and 800 workers on the project and keep them all on track. It is integrally connected to the other chapters in this book—tracking the cost and schedule, planning ahead, and managing people.

First is managing the main office staff and owner of the subcontracting firm. With new subcontractors, the first thing I would always do was travel to their office, see their operation, and establish a relationship. Almost no one in construction does this because somehow we think that they should come to us since we hire them. But that's a mistake.

The owners need to know you care about their operation and are concerned for the success of the project, which includes their making money. It cannot be a one-way street and going to them shows them you understand. I have been told many times that no one ever does this, but they get that it means I want to have a close relationship and will watch everything closely. It also shows you if they are capable of performing.

MANAGING THE TRADES

Paying their office a visit is also a good idea when you are bidding with a new subcontractor and want to see their operation. Otherwise you have no way of knowing if they can take on the work. Their price may be competitive, but you need to know they will succeed. And even more importantly, on bid day if they are very low, you may want to knock them off the list of bidders and protect yourself from taking on a sub that you will have to prop up or, worse, who will fail.

I had a project at the University of Virginia In Charlottesville, Virginia many years ago. We bid blind in a city that we had never worked in before, two hours from Washington, DC. I was running the office and my president was in the room during bid time. In this case we had not vetted any subs and never visited them. The president was aggressive, cut the bid prices from the subs too much, and we won the project as low bidder.

When my project manager and I finally went to Charlottesville, and tried to buy out the subs, we got a rude awakening. It was a cabal and they all knew each other and hated contractors from DC. By the time we went from one sub to another, they had already talked and decided they would not lower their prices at all. Worse, the University facilities head was friends with all the local subcontractors. This was a disaster as we had a bid bond guaranteeing the bid and we were going to lose money. Only through the aggressive pursuit of delays and change orders were we able to come close to breaking even. We had not done our homework. I never forgot that and always made sure I met with the subs at their offices.

Second, and very different, is managing the foreman of the subs and their workers on-site. You will run into all different personalities—the ones who will work with you and be reasonable and the ones who will fight with you every inch of the way. Here again, as soon as they get on the jobsite, you need to figure out who they are and how to get them on your side. It takes time, meeting constantly with them.

Know what makes them tick, what their concerns are, who you know in common, and see them every day. The best way to do that is meet them early, as the trades start as early as 5:30 or 6:00 a.m. When you catch them early, you can get them while they are planning the day and can rearrange their work schedule if needed. Later in the day, by 2:00 p.m. they're going home.

You also need to follow through on what you promise. If you tell them to proceed and you will sign their extra work order later in the day, make sure you do it. The first time you don't, you will never get anything from them again. You also need them to hold to the same rules. If they promise something and they don't do it, make it clear they will pay a price if they try that again. And follow through and punish them if they don't. The trades are a bit like children this way. They need to be trained on every project.

One of my assistants, Phil Kirby, took this to a new level at the Solomon Brothers HQ, and I always wish I had thought of it. The electrician was supposed to move to another floor from where they were working and finish roughing in some studs with cable. The foreman promised it would get done. But this guy was known to not always follow through. So, in the middle of the day, Phil realized that the foreman had not kept up his part of the deal. He was promised overtime and extra work if he did it, and he just didn't care.

Phil went to the lockers of the electrician and took all the workers' clothes out. When they got back at the end of the day to go home—all forty of them: no clothes. The foreman stormed into Phil's office, knowing it was him, and went crazy. Phil looked at him and said, "You get the clothes when you finish the work." The foreman gathered his men and worked overtime to finish the work. Now that is managing the trades.

CHAPTER 30

Know When to Move On

I am not the poster child for creating a career in one firm for your entire life. My father started in high school working for his newspaper in Boston and stayed at the same firm for his entire career- forty years. I never believed in that. I've had fourteen jobs since I graduated from college. That being said, I never changed for money; I always moved on because I wasn't learning anything and saw that I would be stuck in a rut at that firm if I stayed. Luckily, even though construction cycles up and down, I usually hit the market at the right time when I moved.

The first sign that it's time for a change is if your bosses are too young. At HRH Construction most of the executives were only five or ten years older than I was. Unless the company is growing incredibly fast, you will never advance at a pace you will be happy with if that is the case. Once I advanced from assistant project manager to full project manager, I knew I was stuck in that role forever. And I wanted to be an executive at some point. I moved on to Tishman Construction with more responsibility and more money.

I used to complain about money all the time. I was smart enough to never complain about my end-of-year bonus but would always complain about my salary. If you are in a large firm, you are always stuck with your 3 percent increase, which will never get you anywhere.

As I mentioned before in this book, one day I was having lunch with Artie, and he looked at me and said, "I'm going to give you some advice. Don't focus on the money. Be the best at what you do, work harder, achieve more, and the money will follow." I stopped talking and took his advice. I have given this same advice to others many times with mixed success. But Artie was right.

CHAPTER 31

How to Handle the Interview

We all find ourselves sitting across from the person who may or may not hire us and never quite know what to say or if we did a good job. There are many books on this, but my advice is fairly simple. Relax and be yourself. If they don't like what they see, then you probably don't want to work there. They already have your résumé and may know you by reputation or have checked you out. So, here is my advice for interviews:

- Research the firm you are interviewing with and the person or people you will talk to. Find out, if you can, who you will interview with when their assistant sets up the meeting. Make sure you know someone else at the firm and call them to discuss where the firm is at this time, what projects they are working on, why they might be hiring you, and discover the culture of the firm.
- Write down everything the assistant tells you when setting up the interview, including the assistant's name. When you walk into the office, you will most likely face the assistant first. Call them by name.
- Bring paper copies of your résumé. If you don't have it with you and assume the interviewer will have it, you will be embarrassed

when they ask for it. Executives are busy and often are seeing you in a hectic day of meetings. I used to ask applicants for the résumé just to see if they brought it with them.

- Be calm and don't talk too much or interrupt. Let the interviewer set the pace and follow where they are going with the interview. You will try to guess what they are going to ask you, but you will most likely be wrong. You don't know how you might fit into their staff.
- Answer questions the way you want to. Respond to the question but spin it to elaborate on why they should hire you. You can give a quick answer to the question and then take it as an opportunity to sell yourself.
- Be humble but confident. Don't brag or talk about me, me, me. Talk about how you work with teams, collaborate, and manage. They are not hiring you to work in a vacuum.
- Don't start interviewing the interviewer. I had several interviews where the candidate started asking me about my role in the firm, my history, or what the firm was looking for. I actually would say in this case, "I'm interviewing you, not the other way around." Believe me, they didn't get hired after I said that. It was a killer comment.
- When the interviewer asks you what your strengths and weaknesses are, start with the strengths. You know how to sell yourself, your team-building qualities, successful track record, etc. The weakness thing is a trap. Remember when Obama was asked in an interview while running for president in 2008 what his weakness was and he answered, "I'm not very good with paperwork." Really, why would I want a president who can't do paperwork and thus is not organized? My answer, developed over many years, is to say, "I work too hard and such long hours that I feel my family suffers from this." It's a negative cloaked in a compliment to yourself that you work too hard. Question dodged.

- Know your résumé backward and forward. Don't ever look at it in front of them; just have a copy in a folder. If you need to look at it, you will look tentative, as if you don't know your own history.
- Tell a story if you can. If they ask you how you would manage a difficult subcontractor or architect, be prepared with a story showing how you solved the problem. We all like stories.
- Find out who you know who knows your interviewer. Use that as a reference and after the interview call them and see if they can influence your being hired. And never put someone on a list as a reference if you're not sure they will describe you as the most amazing employee they ever hired. I have called references when I was interviewing candidates and actually gotten mixed reviews on them. That is a deal killer when I am looking at many qualified candidates. Let the references know who might call them before the call happens.
- Wear a suit and tie or dress (only if you are a woman). I know the millennials don't like to get dressed up, but it is a sign of respect. The interviewer may tell you that a suit and tie or dress is not needed if you work there, but for the interview it matters. If I saw candidates who were too casually dressed, I took it as their mistake.
- Look them in the eyes—not in a creepy way—and don't glance around the room. Even if you are nervous, if you look tentative, they will assume you can't push a construction project.
- Have references available from subcontractors. Most times when I was interviewed, the interviewer wanted to know that I not only knew architects and engineers or general contractors. They wanted to know if I could get along with the trades.
- Follow up with a note or email thanking them for the time for the interview. It makes a difference, and most people I interviewed didn't do it.
- Don't call twice a week to see if you are being hired. Give it at least a week and then call the assistant. Don't email the

interviewer for at least two weeks. It's annoying to appear to pester them, and it makes them offer you less money if they do offer you a job, as you look desperate.
- Be honest if they ask you how much money you are making now. Lying to increase your salary may backfire; they may already know your salary. Everyone knows everyone in construction. Don't dictate how much you want to make. Let the interviewer discuss it, which most likely they won't in the first interview. Don't take a job offer for less money than you make now unless it's a great opportunity. It makes you look weak.
- Have a clever question ready for them, because they most likely will say at the end, "Do you have any questions for me?" I like the comedian Steven Wright's question: "If a car is traveling at the speed of light and turns on its headlights, does anything happen?" But seriously, ask them something about their firm that will show you did your homework, or about a project if you know why they are hiring you.

CHAPTER 32

Love What You Do or Find Another Career

My father used to say to me that you need to wake up every morning excited about going to work. As a child I didn't understand what he meant until I started working and making my own money. But I've been fortunate to love what I was doing, even from a young age.

This actually started at age ten in suburban Boston when I got my first paper delivery route. It was most likely illegal child labor, but the local distributor of newspapers had an army of young boys and girls riding their bikes around throwing the morning paper onto the doorstep of the many houses in our community. And back then newspapers were the way everyone found out what was happening as they started their day.

We would get our bundle of newspapers delivered to my parents' house at 6 a.m. I would wake up excited to start my daily routine, which would take an hour to accomplish. And I would make five dollars a week, including the Sunday paper, which required me to drag a Radio Flyer red wagon around because the Sunday papers were so heavy. But five dollars a week that I could spend any way I wanted was a fortune. I can't tell you the feeling of being independent and making money that I controlled. I loved that job.

And then into my teenage years. I got a job making deliveries in a flower shop in my hometown. I was working for Freddy the Florist, a local florist in Wellesley, Massachusetts, a sleepy, wealthy community near Boston that is home to the college of the same name. I was introduced to Freddy the Florist (Fred Gamer being his actual name) by a friend who delivered flowers for him. Fred was my answer for a boss—he was irreverent, ADHD, funny, and rewarded hard work with praise and friendship that you couldn't buy. He was the perfect boss.

It was like working for myself. Fred provided support and training where needed and then let you flourish. He taught me how to get things done. Sounds simple but most people spend their entire working life finding ways not to get things done or blaming others for their failure to get things done. Work after all is about getting stuff done. There are no excuses, no matter what obstacles are in your path. And the first thing you learn about having a job is that the world is determined to prevent you from getting things done.

First lesson from Freddy: When you go out to deliver flowers, don't ever come back with undelivered flowers. After all, flowers are perishable. The minute Freddy bought them at the market, the clock was ticking on their imminent death. So, summer or winter (winter actually being worse for flower delivery as freezing kills them instantly), you didn't come back to the shop with deliveries. Many of my friends tried to work for Freddy delivering flowers and got fired. The funny thing is Freddy had a temper and could be very tough. But he never yelled at me. Because I did what he told me to do.

I worked for Freddy all through junior high and high school. Six years. No one else came close to lasting that long with Freddy. And once again I had money in my pocket for whatever I wanted, not to mention free flowers for my mother. No one could tell me how to spend it because it was mine. My mother was a saint and tried to get me to save. But I always knew that life was short, and I loved to buy things. Anything. Consuming pleased me. To this day that is true. It drove my mother crazy, as she was a saver. My wife has an expression that I like: "Don't spend less. Make more."

I only came close to failing Freddy once. It was late one winter night after school. I had a delivery to Doherty's Funeral Home, a big destination obviously for a florist. Now funeral arrangements were particularly time sensitive because we would use all the flowers that were dying, and no customer would want. We would fluff them up, put them in water picks (little plastic tubes filled with water), and then get them to the funeral home as soon as possible. They only had to look good for twenty-four hours and then they were trash. A florist's dream.

It was 6:30 at night, last delivery, the flower shop was closing, and I just had to deliver one arrangement and return the truck to the shop. I had my own set of keys and Freddy was gone.

I turned in to the dark and creepy parking lot of Doherty's and parked. I knocked on the front door, where there was usually an attendant on duty because wakes of course were at night. This time there was no one there. Panic set in. I couldn't tell Freddy I failed. I had a perfect track record. As I was about to give up, I remembered there was a cellar door in the back. Maybe it was open. Remember, I could not leave the flowers outside because they would freeze.

Only one problem—the cellar was where they embalmed the dead bodies, and I would have to traverse the lab to get upstairs. An impossible choice: fail and go home or walk through the Valley of Death. I chose the Valley. I remember opening the cellar door, entering slowly so as not to bump into any bodies and looking for a path to the upstairs. I had never been in the cellar. I saw a light shining from upstairs—my roadmap. I set my sights on the light at the bottom of the stair and didn't look right or left. I knew there were dead people there and could smell the embalming fluid.

I made my way upstairs, placed the flowers in the vestibule, and luckily could then leave by the front door and lock it behind me. No alarm system needed in a funeral home.

When I got outside, I was a different person. I had looked death in the face, and death had blinked first. It felt good to know I still had a perfect record. And I loved Freddy as a boss and had not failed him. The next day I told him the story, and he laughed for twenty minutes. I told him it was not funny, but that made him laugh even more.

Freddy also taught me to love my work, but ultimately it was in my personality to want to get up every day and do something meaningful. It was my fear of failure. Freddy and later Artie Nusbaum showed me the way. Construction is the greatest career anyone could wish for. I started out by saying how lucky I was to find myself at the young age of thirty building a sixty-five-story tower in Manhattan. I hope you find the same life in construction and have a winning career.

www.ingramcontent.com/pod-product-compliance
Lightning Source LLC
Chambersburg PA
CBHW050115230526
45470CB00004B/1844